Wildcrafting For Beginners

Step By Step Blueprint To Wildcrafting Made Easy For Complete Beginners

Introduction

Do you enjoy foraging or the idea of getting food and medicine freely from mother nature?

Would you love to know how best to identify different plants to ensure that the plants you harvest are fit for human consumption but are not sure where to get such comprehensive information?

If yes, you are in safe hands. This book will give you all the information you need on wildcrafting as a complete beginner so that you can become an expert wildcrafter!

The truth is, wildcrafting can be intimidating at first when you don't have enough knowledge of plants and are searching for edible plants in the wild, plants great for medicines and poisonous plants that you need to avoid.

However, like any other skill, once you master it, you will be surprised at how naturally it will come to you. This guide will bring you to that level of comfort in wildcrafting if followed correctly.

Read along as we will answer all your questions, including:

- What exactly wildcrafting is

- Where wildcrafting came from, and why learning such a skill is beneficial

- How to correctly identify plants

- How to be ethical and sustainable in wildcrafting

- Some principles to follow when harvesting plants

- What should you do after the harvest

- Tools needed

- Some standard outdoor safety protocols you need to follow

- And much more!

Let's dive right in!

Table of Contents

Chapter 1: Wild-crafting and Foraging

Have you ever noticed some mushrooms growing in natural environments such as forests and bushes?

Did you pick them?

If you did, you are already a wildcrafter/forager!

So what exactly is wildcrafting?

Wildcrafting is an ancient practice that fell out with the introduction of modern pharmaceuticals and doctor visits. In traditional medicine, men used herbs obtained by wildcrafting, and herbalists had the skill and knowledge to know which plant could treat what illness, and this book will pass part of that knowledge to you.

Basically, wildcrafting is the practice of gathering plants, whether flowers, roots, herbs, nuts, seeds, or even leaves from wild sources such as forests, to be used as food, natural medicine, or cosmetic treatment.

You might have also come across resources referring to wildcrafting as foraging. It is almost the same; the difference is that foraging refers to gathering food, while wildcrafting gathers plants for food, medicinal purposes, and beauty treatments.

Human beings were meant to co-exist with plants; therefore, we are part of the same ecosystem. There exists some balance between us humans and plants.

However, while plants can heal, others can poison. That is why having good knowledge and understanding of plants is essential so that you do not end up consuming toxic plants or missing out on the value that some plants can add to you.

So, What Are Some of the Benefits of Wildcrafting?

Wildcrafting has a lot of benefits, including:

1. An opportunity to enjoy the outdoors and nature, alone or with the great company of friends or family. Wildcrafting is fun, especially when you do it with friends. In addition, spending time outdoors in nature wildcrafting or just wandering in the wilderness has its own health benefits.

   Studies show that those who spend more time outdoors are likely to live longer. The lengthy lifespan may be due to factors such as spending time in the natural world, which exposes you to clean air and gives you good exposure to vitamin D. Also, reduced stress and anxiety increase happiness and contentment.

2. While you are out wildcrafting, you get to harvest some of nature's wildest herbs which can be used to make herbal remedies at home to treat various illnesses. This can reduce dependence on imported medicine.

3. Wild-crafted plants have more nutrients because they grow in richer terrains that have been less disturbed than plants growing in commercial farms that feed on isolated nutrients.

4. Foraging and wildcrafting are ways to learn about ecology and plants' natural habitats. The more you interact with natural environments, the better you understand them. It will also save you money because everything out there in nature is free yet of immense value to your health.

Chapter 2: Rules/Principles/Ethics of Wild-crafting.

Before we get started with wildcrafting, there are some rules and principles you need to know before going out wild-crafting.

Let us learn more about those below:

1. Get permission

Be sure to ask for permission before harvesting any plants. Not all public land can just be accessed, and even if you can access it, the plants may be part of some projects and may not be suitable for you. If it's private land, ensure that you get permission from the owner first.

Also, you need to know the traditions, so you do not go against them. For example, some people have followed practices concerning foraging that were adopted from their ancestors, and it is vital that you respect them.

In some traditions, the ancestral natives of the land respected seasons and harvests. For example, certain plants are ready in certain months of the year, and harvest must happen only in those months, while some other traditions give thanks after harvesting. Some of these traditions might still be in practice, and you need to know them too.

2. Be entirely sure of the plants you are harvesting

Some plants look the same but are very different. Their use could be very dissimilar, while others could be harmful. Carry a guidebook or pictures with you for accurate identification of the plants. You also need to know if the plants are perennial or millennial to have continuity in mind even as you harvest.

3. Do not go for the first patch you see

Chances are, everybody who walks into that field sees that same patch. Instead, go further deeper so that you get a fresh patch. Avoid patches near industrial areas and under power lines, as the plants may be polluted with toxic runoffs. Also, patches next to the roadside and near commercial farms may be contaminated with pesticides.

4. Stay safe

Wear the right clothes and shoes, and be aware of your surroundings. Watch out for low-hanging branches and anything else that may be a hazard to you while you are away from the rest of the world.

5. Do not trash the environment

Don't leave trash that would damage the plants. You want to ensure you still have plants growing in the future. Fill the root holes so the next person will safely harvest in the same place.

Generally, leave the environment better than you found it.

6. Follow foraging rules

Leave more plants than you take so that there will be some left for future harvests if you want to keep enjoying the benefits of nature. Therefore, only pick what you need.

This is unlike if you harvest a lot only to waste the plants because they are more than you need. For every ten plants, harvest one; whenever you can, plant an extra seed of the same variety you harvested.

7. Protect the vulnerable plant population

Plants that are rare and unusual and are going extinct need protection – so avoid picking them. If you can, nurture them to multiply them.

8. Always carry proper tools that will not damage the plants

The goal is not to distract the growth of the plants. Therefore, proper tools will help cut only the part of the plant you want, leaving the rest to blossom. In a later chapter, we will mention the tools you need as a beginner for wildcrafting.

9. Assess your area first

Avoid areas that may have had pesticides used on them to avoid toxins. Go at least 200 meters from the road, where you can be almost sure no one uses pesticides on the plants.

10. Offer gratitude to mother nature for sustaining the plants and respect mother nature.

Have a gratitude ritual for the gift of those plants.

Chapter 3: Preparation for Wild-Crafting- Tools Needed and Safety Protocols

To prepare yourself for the wild, you will need tools that will be essential in your excursion.

1. A basket, paper bag, or cloth bag

You will need a place to put your plants and herbs.

Paper bags are good for long-distance trips but can decompose the plants if left there for a long. However, they are good because they can carry large harvests.

Mesh bags are also a wise choice as they let your plants breathe and allow air circulation.

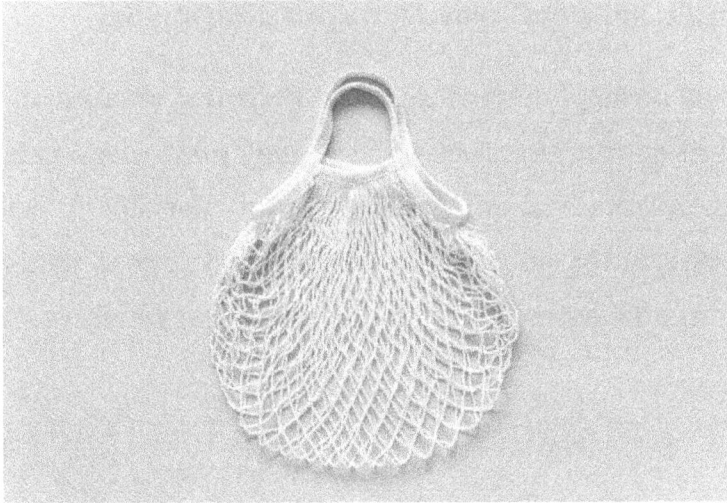

Baskets, especially the open ones, will allow you to use both of your hands to harvest because they can rest on the ground, and you just drop the plants in.

You can choose either of the above options depending on your needs.

2. Knife, scissors, trowel, or pruning shears

A **knife** will be helpful in cutting bark. **Pruning shears** will be useful in cutting branches because they make clean cuts that let the plants heal efficiently. It is not advisable to cut the branches using your hands as doing so may expose plants to infections and diseases through the rough scar you leave.

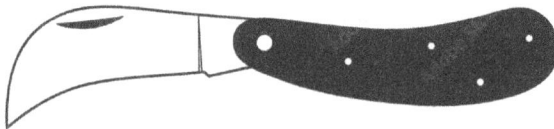

Scissors or sickles can be used for snipping flowers, and a **trowel** for digging roots will also be useful so that you do not slice the roots on the ground.

Hori Hori is a multipurpose tool that can be used for cutting, pruning, and digging. It has a serrated blade, a straight blade, and a curved blade. If you have it, you will need no other tool.

You will need these tools depending on the kind of harvest you are looking for.

3. Vegetable brush and a mushroom brush

A vegetable brush is another handy tool that you will need if you do not want to carry a lot of dirt with you. The brush is useful for dusting off dirt from the roots.

A vegetable brush might not be the best for mushrooms as it may damage them. To avoid this, there is a special brush for mushrooms (see the last image above).

4. Snacks and water

You are likely to take long hours out there in the wild. Some snacks to give you energy and water to keep you hydrated will go a long way.

A sandwich, some bananas, and fruit will do. Do not rely on the water you find in the wild, as it may be contaminated, and the goal is to get home safely.

5. What to Wear

The first thing to check when deciding what to wear is the weather, then dress accordingly. Regardless of the kind of weather, though, long pants and long sleeves are the best for wildcrafting because they protect your skin underneath from plants like poison ivy that can cause a skin rash.

The terrain of the area is another thing you need to factor in when choosing the kind of shoes to wear. Hiking boots are the best because they are waterproof and cannot easily be penetrated by thorns.

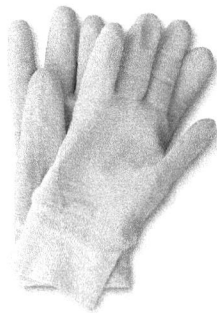

Gardening gloves are not always necessary but will be super helpful in protecting you against thorns and plants that cause skin irritation, such as stinging nettles.

Don't forget your hat, as it makes a big difference in the scorching sun.

6. Record sheet

Do you have a checklist of particular plants you would like to find? No problem whether you have one or not. If you are one of those people who love to keep records, a record sheet will help you make a list of the plants you have identified, and the list could come in handy in the future.

Safety Protocols When Wildcrafting

Wildcrafting involves sometimes being in untapped areas, wild areas, and forests, and we know this is home to most undomesticated creatures that might not always be friendly to people. That said, it is crucial to be aware of the area you will be visiting.

Below are a few points to note:

If you must go wildcrafting alone, let somebody know where you are going. If you can, send your pin to a loved one so that you can be traced in case of anything – most smartphones have maps that can help with this. Make sure to report back when you arrive back home.

A camera and a plant guide are essential for properly identifying plants. How is this a safety precaution? You may ask.

For instance, all mushrooms may look attractive, and you may be tempted to quickly run for them thinking it is your lucky day, but not all mushrooms are edible. Some are very poisonous, and poisoning yourself or others is not part of the plan here. A guide will help you identify the right plants.

When harvesting plants with a toxic look alike, ensure that you have a clear picture of both, so you don't get confused by their similarity.

Don't harvest if you are in doubt. Leave it for another day when you have confirmation. A camera, in this case, will help to take a picture for reference later.

You can take your dog along with you so that its sharp eyes and ears will help alert you of bears and other predators around you.

A can of pepper spray is a good weapon when you are out there in the woods.

A first aid kit is also vital in case of any injuries that might occur while you are in the wild.

Ensure that you are going to the right place. From a legal perspective, in some countries, it is illegal to disturb, harm, or cut down plants in specific areas like forests and parks. So ensure that you harvest in the right places. Also, avoid roadsides and old landfills and go deeper into the land where plants are clean and just natural.

Use insect repellant cream to protect yourself from tick bites and other disease-causing insects like mosquitos found in the woods.

Beware of seasons. It tends to get slippery when it rains, so you want to choose a time when the ground is dry and the plants are not too wet as well.

Chapter 4: Plant Identification

Wildcrafting may come off as an easy task, but it's a skill that will take you time to learn. The same way you observe a person, their characteristics, and their features to know them is the same in plant identification. You pay attention to all the plant's features and details for you to know it thoroughly.

Observation will be one of the virtues needed in identifying the plants. Indeed, take your time, get acquainted with the plants, and soon, it will be a -no-brainer for you.

The secret is to start small. What does that mean? You can begin by observing and identifying a single plant and then slowly identify another one, and soon you will be sure of the plants you need to pick.

Start by taking note of plants available in a particular habitat. Before you can even start mastering them, take note of the neighboring plants growing around, notice the ground and the type of soil it grows on, water bodies around the area, if there is sunlight reaching the plant, and what the plant looks like. All these factors, the habitat, and the immediate surrounding environment may reveal a lot about the characteristics and features of a plant. This will be an excellent first step to mastering plants.

Learn Plant Families, Habitats, and Seasons

One of the ways you can identify plants is by knowing and observing their families. Botanists created groups with the same patterns and called them families. Plants in the same family tend to have similar patterns. For example, mustard and mint families worldwide have different appearances, but they have patterns that are alike across them all.

Take note of the habitats for particular plant species to grow. Is it near water bodies, on other trees, on the ground, and the type of soil they grow in. This will help you save time, as you will know where to find them.

Observe the plants through different seasons to see how they change through the seasons.

In addition, learn the terms used to describe plants and different parts of plants. You will notice plants are named by their biological names.

Regardless, plant features and details are more important to know, and the names will sink in with time as you keep referring to the guidebook and interacting with the plants.

Standard Terms Used When Identifying Flowers

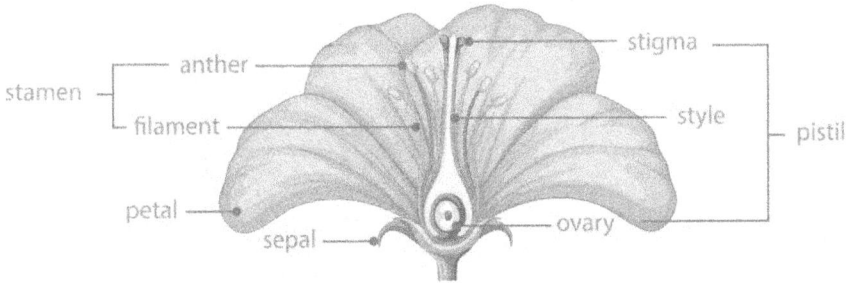

Petals are modified leaves surrounding the reproductive parts of a flower. Usually, they are colored. Refer to the above image.

The pistil is the female part of the flower with the stigma, style, and ovary. A pistil has several carpels.

Sepal is mostly found at the bottom of the flower, and it is meant to protect the flower bud. It is mostly green but can be a different color in different plants.

The stamen is the male part of the flower that comprises the long tubular filament and a sac at the top called the anther.

Regular vs. Irregular flower- A regular flower has identical individual parts like the color, the petals, and the sepals, while an irregular flower has different individual components.

All about Plant Leaves

a) Parts of a Leaf

Below is a picture showing all the essential parts of a leaf:

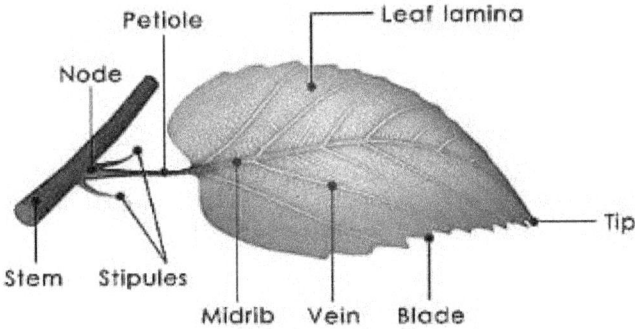

b) Arrangement of leaves along the stem

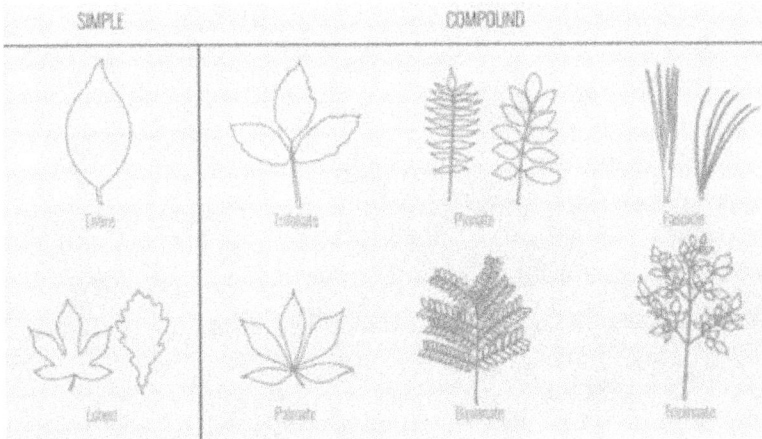

Opposite- Leaves are attached along the stem in pairs opposite of each other.

Alternate- Leaves are attached along the stem in an alternating pattern.

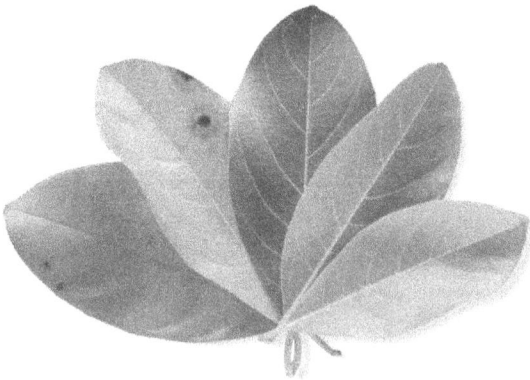

Wholed –The leaves originate from a single point on the stem.

Pinnate leaves –Leaves are divided into leaflets, and there is an even number of leaflets on a twig.

Palmate- These are compound leaves that have a fan shape.

Types of Leaves

Rounded –These are leaves that have a round shape

Ovate- These are leaves that have an oval shape

Cordate- These are leaves that have a heart shape.

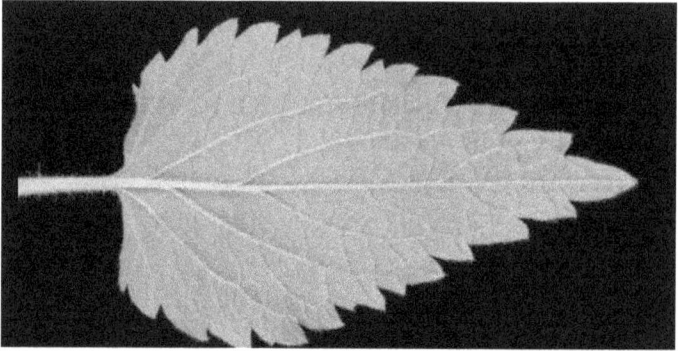

Sagittate-This type of leaf has a triangular shape at the base and two acute lobes at the head. (arrow-shaped head)

c) Leaf Margins

Serrated or toothed –leaves that have sharp teeth

Entire – leaves that have smooth margins with no teeth.

Dentate – These types of leaves have rectangular or square teeth along the leave margins

dentate

Double Dentate – this is a type of leave margin that is toothed, with each tooth having smaller teeth or serrations.

Lobed Leaf margins – these are leaves with deeply indented margins, e.g., the pawpaw tree leaf.

lobed

Keep a sketchbook where you can draw the plants you have identified. Include all the details of the plant in the book. This way, you will internalize the plant features quickly, and identifying the plants will be even easier.

Take pictures of the plants you have identified and organize the photos in name albums. You can use a camera of your choice; a phone camera will also work perfectly. When identifying plants using the pictures, you can zoom in for tinier details that are not easily seen.

Use plant identification apps on your smartphone such as:

Plantsnap:
(https://play.google.com/store/apps/details?id=com.fws.plantsnap2&hl=en&gl=US)

Plant Identifier

(https://apps.apple.com/ca/app/plant-identifier-picture-tree/id1567330581)

ID weed

(https://play.google.com/store/apps/details?id=com.extension.idweeds&hl=en&gl=US)

The downside of using apps is that they can be unreliable, especially in areas with a weak cell signal, because you might not be able to access the app. The other downside is that you will not learn since the app does everything for you. **Plant Families**

Let us look at some identifying features in specific plant species and families.

1. The Mint Family

The plants in the mint family have a square stalk, simple opposite leaves, and small flowers and they are aromatic. The scent is essential in your identification.

Plants in this family have a spicy quality that is attractive for cooking. Other plants in the same family include rosemary, lavender, marjoram, savory, thyme, peppermint, lemon balm, and sage.

Many plants with square stems and opposite leaves can easily be confused with the mint family. These do not have that aromatic smell and belong to the stinging nettle family, Loosestrife, and verbena family.

Familiarize yourself with the flowers as well because they have very similar features. You will notice that they have four stamens and five united sepals, and their flowers mature into a seed capsule containing four nutlets.

MINTS FAMILY PLANTS

Plants in the mint family grow in moist soils and wet environments.

2. Identifying Plants of the Parsley Or The Carrot Family

You should be keen to identify the carrot family because it has some toxic look-alikes that can easily confuse you.

The carrot family, also known as the *Apiaceae* or the parsley family, has members like coriander, parsley, dill, fennel, and celery. These are among the 1500 members available worldwide. The hemlock and the hemlock water dropwort is the toxic look-alike of the carrot family to watch out for.

Some of the specific features of the carrot family include:

- Ribbed and hollow stems.

- Compound leaves, normally alternate and have a generally sheathed stalk at the base.

- Flowers are simple or compound umbels, with five petals normally uncurved at the tip.

- Flowers are tiny and are either white or yellow.

- Their roots thicken and lengthen as the plants grow to allow access to deep water reserves. Starches, sugars, and other nutrients are stored here.

The poison hemlock plant resembles the carrot family but lacks hairs on the stem, and this is one of the distinguishing features and the purple-reddish blotches on the stems.

3. Identifying the pea family

If you have grown beans or have seen beans growing, it will be easier for you to identify plants in the pea family. The pea family is also called the legume family or *Fabaceae*. This family includes plants like beans, alfalfa, lentils, soybeans, peanuts, peas, and clover, among other 13000 species worldwide.

Some of the iconic features of plants in this family are:

- Compound leaves, often pinnate and generally alternate – compound leaves are leaves that have leaflets on a common axis.

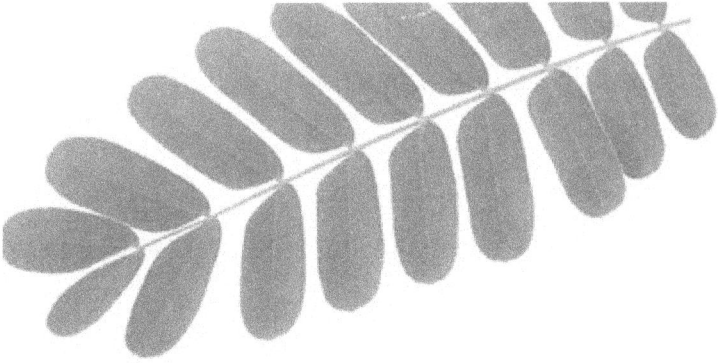

- The flowers in the pea family in different subfamilies have unique floral designs, but mostly the flowers have five petals in an arrangement of 2 banners, 2 wings, and 2 keels.

- The fruits in the pea family are the legumes customarily arranged in a pod closed in 2 seams.

For practice, observe the flowers and notice the banner, wings, and keel.

4. Identifying Plants in The Sunflower Family

The sunflower family, also known as the Daisy family or the *Asteraceae*, is one of the most cheerful and colorful flower species. It is also one of the largest flower species, with about 23,000 species worldwide.

The sunflower family has iconic features such as:

- Simple to alternate compound leaves. A simple leaf appears alone and does not have leaflets on the axis.

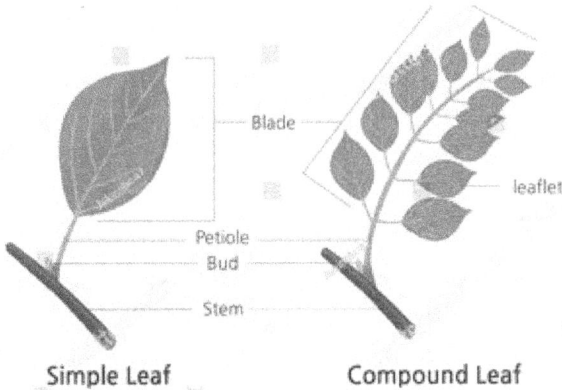

| Simple Leaf | Compound Leaf |

(Labels: Blade, leaflet, Petiole, Bud, Stem)

- They are annual and herbaceous perennials and can also be herbs, shrubs, and trees. The annual plants are those plants that grow for only one season and then die off, while the perennial plants are those plants that bloom for only one season a year.

- Their flowers have a larger head with hundreds of individual flowers. There are four basic types of these flowers.

  The radiant head has a ray of flowers and a disk with 5 or 3 lobed star flower clusters in the center or a flower rim at the edge, as seen in the mule ears.

  Disciform heads have disk-like flowers with 5 lobed star flower clusters in the center or disk-like flowers with missing ray rims on the edge, e.g., the everlasting.

Discoid heads have 5 lobed star disk flowers, such as in the mugwort.

Ligulate heads have 5 lobed strap-like flowers; for example, the dandelion.

The sunflower family is divided into sun families according to their sizes and similarities. Some of those subfamilies include the Mayweed tribe, which has flowers like the yarrow, sagebrush, and pineapple weed. The sunflower tribe has flowers like goldfields, tarweed, and mule ears.

In the next chapter, we will focus more on the benefits of these flowers, whether they are edible, herbal or have cosmetic benefits

5. Identifying Plants in the Mallow Family

If you know the hibiscus flower, you know what kind of plants are found in the Mallow family. The biological name for the mallow family is *Malvaceae*. This family includes plants like hibiscus, cheeseweed, flannel bush, and cotton, among others.

Cotton exhibits poisonous traits, and it's the only known toxic plant in that family and is useful in other ways.

The wild ones may be smaller, but the standard iconic features across this family are:

- Broad leaves, alternate, usually palmately lobed and toothed with star-like hairs.

- Funnel-like flowers that have five separate petals and a distinct column of stamen surrounding the pistil.

- The mallow family is mostly found in warm climates; worldwide, there are around 4000 species.

6. Identifying the mustard family

If you already know what cabbage or broccoli looks like, it will be easy for you to understand and identify plants in the mustard family. The biological name for the mustard family is *Brassicaceae.*

There are about 3700 species in the world, and some of these species include turnips, bittercress, cabbage and its relatives, peppergrass, rocket, sea kale, and white mustard.

Some of the iconic features of the mustard plant family are:

- They are usually herbaceous perennials, shrubs, and annuals. That is, some grow for only one season while others bloom yearly.

- The leaves are both basal and alternate and are generally simple.

- Flowers have four petals that form a cross. The stem is unbranched.

7. Identifying grass family

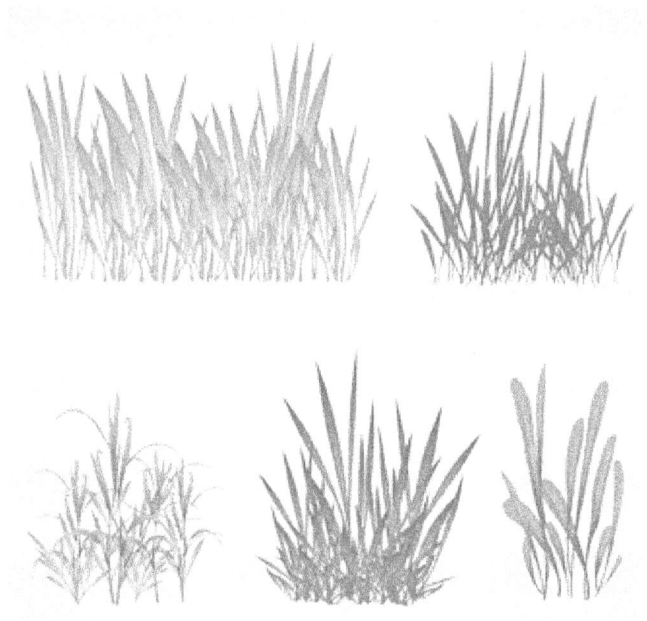

This plant family has plants such as maize, millet, barley, and wheat, which are pretty common. That said, learning and identifying plants in the grass family will be easier because they all have pretty similar traits.

The biological name for the grass family is *Poaceae* or *Gramineae*. Grass family members are monocotyledonous flowering plants, including cereal grasses, bamboo grasses, and natural grassland grasses.

There are approximately 10,550 grass species worldwide. The plants in the grass family are maize, millet, rice, barley, rye, fescue, brome, and bamboo, among many others.

Some of the most common features of the grass family include:

- A hollow stem is most significantly seen in the bamboo tree.

- They mostly have narrow, sheathing leaves, and they are alternate. Sheathing leaves are those leaves that have a wide base that surrounds the stem as an envelope.

- They have small petalless flowers, and the fruits are mostly grains, as seen in wheat and maize.

8. Identifying the miner's lettuce family

Miner's lettuce is a short green plant family also known scientifically as *Montiaceae*. There are approximately 230 species across the world, and they include plants such as pussy paws, red maids, and bitter roots, among others.

Miner's lettuce mainly grows in the coastal and central plains.

Their common features include:

- They are herbaceous, annual, or perennial plants.

- They have simple leaves, fresh and often succulent to adapt to dry conditions.

- The leaves may be alternate or opposite.

- Their flowers have 2 sepals, with a superior ovary.

- They are mildly acidic with some oxalic acid.

9. Identifying The Plantain Family

You may know plantain to be a banana-like fruit, but in this case, that fruit is unrelated to the plantain plant family. Bananas and plantain belong to the *Musaceae* family.

Plants in the plantain family are also scientifically known as *Plantaginaceae*. These include plants like snapdragons, hyssops, and speedwells.

Some shared features in this plant family include:

- Simple leaves which can be alternate or opposite and are sometimes parallel.

- They have florescent flowers, often two-lipped, and with superior ovaries.

- The plantain family mostly grows in temperate zones.

- The plants have a taproot system.

Identifying Oak Family

Oak is a family of trees and shrubs. Their scientific name is *Fagaceae.*

Oaks grow in a wide range of habitats, from Mediterranean semi-desert to subtropical rainforests. There are approximately 900 species spread out across the world.

Plants in the oaks family include oaks, chinquapins, beeches, and chestnuts.

The iconic features commonly found in oaks are:

- They are deciduous evergreen trees and shrubs

- The leaves are simple with smooth margins and are usually alternate.

- The oaks flowers have separate male and female flowers on the same plant.

- The fruit is a nut, partly enclosed in a scally cast, as in the image below.

Blue oak tree

11. Identifying the Rose family

Rose flowers are quite common, and you might know them already. The rose family has several other members, with approximately 3000 species spread across the world.

The scientific name of roses is *Rosaceae*. Other plants in the same family include plums, apples, raspberries, toyon, chamise, holy leaved cherry, and ocean spray, among many other plants.

Some of the standard features of the rose family include:

- They are herbaceous, perennials, shrubs, and small trees.

- The leaves are simple or compound, often with serrated edges, and are generally alternate.

- The flowers usually have 5 petals and 5 sepals attached to the cup.

- Their fruits are in many kinds, dry or fleshy fruits.

12. Identifying the Morning Glory Family

Ever heard of the morning glory flower? It is part of the morning glory family. Scientifically, it is called *Convolvulaceae*. The morning glory family has over 1,600 species spread across the world.

The plants are common in warm temperate regions and the tropics. Some members of the morning glory family include sweet potatoes, bindweeds, morning glory, and dodders, among others.

Some of the iconic features of the morning glory family include:

- Funnel-shaped colorful flowers, usually with bright colors. The flowers have 5 fused petals and 5 stamens attached to the petals.

- The leaves are simple with an arrowhead shape, alternate, and lack stipules.

- Most of the plants in the family have stems containing milky sap, most often twinning.

- They are mostly herbaceous vines, perennials, annuals, and some tropical species, including some trees.

- Some species have toxic seeds containing ergoline alkaloids.

- Some weeds in this family are agricultural pests that invade agricultural grounds and crops. Such plants include bindweed. See the image below.

13. Identifying the Iris family

The Iris family is a family of flowering plants and is most diversified in Africa. It has 2,050 species worldwide, and it is scientifically known as *Iridaceae*. The Iris family especially grows in habitats with long dry, and cold periods.

Members of the iris family include plants like the irises, gladiolas, crocuses, and freesias.

Some of the iconic features of plants in the Iris family include:

- The plants are perennial herbs, meaning they live more than two years.

- Plants in this family are geophytes. That is, they have underground storage like the one found in onions.

- The leaves are simple basal leaves, sword-like with a fold in the middle, and the arrangement is generally alternate. Some have a sheathing stem with a fan-like structure.

- The flowers are in small groups or single, and they emerge from the modified leaves. See the image above. You will also notice that the flowers have 3 petals and 3 colorful sepals that are the outer part of the flower, and 3 stamens.

Since the iris family consists of geophytes that store water and nutrients underground, the plant above the ground dies out after flowering and then regrows and survives on the sored nutrients.

Also, for this reason, these plans can survive harsh summer weather and long droughts.

14. Identifying the Phlox plant family

You may have seen the above flowers in someone's backyard or a flower garden somewhere. They're from the phlox family, scientifically known as *Polemoniaceae*.

Plants in this family are flowering plants and usually small flowers. Worldwide, about 314 species are growing, mostly in North America and South America. Plants in this family include phloxes, scarlet gilia, and Western Jacob's ladder.

The iconic features in the phlox family are:

- They are either annual or perennial herbs, small shrubs, and vines.

- They have simple or compound leaves arranged alternately or opposite each other and are narrow.

- The flower arrangement is in many forms, and their parts are in fives. 5 petals, 5 sepals, 5 stamens, and a single pistil.

15. Identifying the Heath Family

The Heath family, Blueberry family or scientifically *Ericaceae,* is a large group of flowering plants with over 3,000 species worldwide. Other plants in this family include blueberries, heathers, manzanitas, rhododendrons, and azaleas.

Plants in this family grow on acidic, sandy, and nutrient-poor soils.

Some of the features found in plants of the heath family:

- They are usually either herbaceous perennials, trees, or shrubs.

Have you ever seen a tree whose bark peels? It was probably from the heath family. Trees in the health or blueberry family have peeling barks. See the image below.

- The leaves are simple and evergreen with a leathery texture.

- Their flowers are in many forms. Often urn-shaped or bell-shaped (refer to the image above) with fused or unfused petals.

- The fruits are usually berries; that is, they are fleshy fruit with many seeds inside, a drupe, which is a fleshy fruit with one seed inside a hard shell, or a capsule which is a dry fruit with chambers that split open at maturity.

A strawberry fruit

Rhododendron fruit

Mulberry

Jabuticaba fruit

16. Identifying plants in the Buckthorn Family

Plants in the Buckthorn family can be easily confused with plants in the coffee family because of the similarities in the appearance of their fruits. Coffee belongs to a family called coffee, madder, or bedstraw family, which we shall be looking at next, but first, let's focus on the buckthorn family.

The buckthorn family is scientifically known as *Rhamnaceae* and has about 950 species worldwide of woody shrubs and small trees.

Plants in this family include California coffeeberry, buckbrush, shiny bush buckthorn, raisin tree, and jujube.

California coffeeberry

Buckbrush

Shiny bush buckthorn

Plants in the buckthorn family are found in temperate and subtropical climates across the world.

Some of the features of plants in the buckthorn family include:

- Usually, they are woody shrubs and short trees.

- They have simple leaves that have curved veins and are pinnate, which means that the leaves alternate along a common axis as in a feather.

- The flowers in plants of the buckthorn family appear in many forms, mostly in dense clusters of small to tiny flowers that have flower parts in fours or fives.

The fruits are either drupe (fleshy fruit with one seed) or capsule (a fruit with multiple chambers containing many seeds and splits open at maturity).

17. Identifying plants in the Madder Family

The madder family is also called the coffee or bedstraw family; scientifically, it's called *Rubiaceae*. It has a lot of similarities with the buckthorn family, which we just discussed above. The coffee we drink comes from the coffee plant of this family.

The madder family is broad and has a variety of plants, from trees and shrubs, to herbs and lianas (a category of plants with roots above the ground and uses other trees for support).

There are about 13,500 species in the madder family worldwide, and some of its members are plants like coffee species, bedstraws, cinchona that produce quinine, and Rubia, among others.

Some of the features of these plants are:

- They have simple and entire leaves (the leaves have smooth margins that are usually opposite or whorled, i.e. having 3 or more leaves at a stem junction), while sometimes the stems can be square.

- Some are annual and perennial herbs and shrubs while others are trees or vines.

- The flowers are in different forms, and mostly, the flowers are small and star-shaped.

- Their fruits are drupes, berries, or nutlets.

18.Identifying the Gourd Plant Family

If you have eaten or seen a cucumber, a butternut, a pumpkin, its leaves, or any of the fruits in the above image, identifying plants in the gourd family will not be too daunting.

Scientifically this plant family is called *Cucurbitaceae,* and worldwide it has about 975 species of food and ornamental plants. Some of the family members are plants like watermelon, cucumber, pumpkin, squash, chayote, thorny melons, muskmelons, and honeydew, among others.

Some caution should be taken when identifying plants in the gourd family because while most of them are edible, some are toxic. For example, the red bryony shown below can lead be deadly if consumed.

Some of the features found in plants in the gourd family include:

- They are herbaceous vines with tendrils.

- The leaves are simple and palmately lobed. i.e., the lobs originate from a single point. The leaves are also generally alternate.

- Their flowers have the male and female flowers separately on the same flower and will often appear in clusters of the male and female flowers at the axis.

- You will also notice that they have 5 fused petals and 5 fused sepals, and the flowers may be funnel-shaped or star-shaped.

- The fruits of plants in the gourd family are a gourd-like berry or a capsule, sometimes with a hard outer shell or cover.

19.Identifying plants in the brake/ fern family

Scientifically the brake family is called Pteridaceae. You are likely to identify plants in this family among rocks. There are about 500 species worldwide, including plants like maiden hair, button fern, sickle fern, and lace fern.

Some of the identifying characteristics of the brake family are:

- They are perennial herbs that grow in rocky places like cliff brake ferns.

- Their leaves are almost always compound with pinnate leaves and have generally dark petioles.

- They have spores on the underside of the leaves called sporangia that occur in clusters called sori on the leave margins. At maturity, the sporangia open, and the microscopic spores are dispersed by the wind.

20.   Identifying Plants in the Buckwheat Family

The buckwheat family is called the Polygonaceae and has about 1,200 species, including plants like docks, sorrels, buckwheat, rhubarb, swamp smartweed, bushy knotweed, and others.

This is one of the plant families to be most cautious about because many of its species are poisonous due to the tannic and oxalic acids. For example, the rhubarb tree has an edible stem but poisonous leaves that have an irritating oxalic acid, which, if consumed, can damage the kidneys.

The stem, however, can be cooked for its rich, tart flavor. In a later chapter, we shall look at all the toxic plants to watch out for.

The rhubarb plant

Some of the features of plants of the buckwheat family include:

- They are mostly herbaceous annuals, perennials, and shrubs.

- They have swollen nodes on the stems with a scarious sheath around the stem called Ochrea.

- The leaves are simple and entire, and each is attached to the Ochrea in an alternating arrangement.

- The flowers are in different forms of clusters held in cuplike bracts. The flowers have 2-3 petals and 2-3 sepals in two separate whorls.

- The fruits are achene, that is, a single-seeded dry fruit attached to the pericarp at one point only.

21. Identifying Plants in the Agave Plant Family

The sisal plant that is used to make rope comes from the agave family, and most plants in the family are similar to the sisal plant in appearance. Therefore, if you are familiar with the sisal plant, you are very close to becoming an expert at identifying plants in the agave family.

The aloe vera plant is also very similar in appearance to the agave family, but they are not related. Aloe vera is from the *Asphodelaceae* family, which we shall focus on next.

Scientifically these plants are called *Asparagaceae*. Another name for this family is the century family. There are about 637 species worldwide. The plants are mostly found in dry and desert areas. Members of the agave family are plants like agave, Joshua tree, yucca, agave attenuate, and others.

Features of plants in the Agave family include:

- Plants in this family are perennial herbs, shrubs, and trees.

- They are geophytes, which means they store water and nutrients underground on the bulb. They adapt well to fire, harsh summer and Mediterranean weather, and extended droughts.

  You will notice that these plants are mostly found in habitats that experience long dry seasons; hence, the stored nutrients help them regrow back after flowering.

- The leaves are usually long, simple narrow leaves, mostly succulent tough fibrous leaves.

- The flowers are a long pinnacle that branches from the stem up. They have 3 petals, 3 sepals, and two separate whorls

- The fruits are dry capsules and multichambered inside, which split open at maturity.

22. Identifying Plants in the Asphodelaceae Family

Aloe vera is one of the most known plants in the *Asphodelaceae* plant family, which is similar to the agave plant family we just looked at. Right after this, we shall look at what makes the two plant families similar but very different simultaneously.

The Asphodelaceae plants are flowering plants with over 900 species worldwide, mostly found in tropics and temperate habitats. Plants in this family have been widely cultivated for ornamentals, and plants like aloe vera are grown for their sap which is used as medicine and for beauty.

Some of the plants in this family are plants:

- Aloe vera

- Gasteria

- Bulbine wolf

- Pasithea and more

Features of the Asphodelaceae plant family include:

- Plants in this family are perennial herbs, shrubs, and trees.

- The leaves are usually long, narrow simple leaves, mostly succulent and easily breakable with a sharp tooth at the tip.

The flowers are on leafless stalks and have woody tissue at the base.

The Asphodelaceae and the agave family are drought-tolerant and adapted to grow in dry and hot climates.

Some of the differences include:

The leaves of the agave family are fibrous and have sharp teeth on their edges, while the Asphodelaceae leaves are fleshy, thick, and filled with sap or a clear gel. The Asphodelaceae leaves are breakable, unlike the agave leaves that have fiber, making them hard to break.

Plants in the agave family grow tall, 1 ft to 20 ft, while the Asphodelaceae family grow just a few feet tall.

23. Identifying plants in the Broomrape plant family

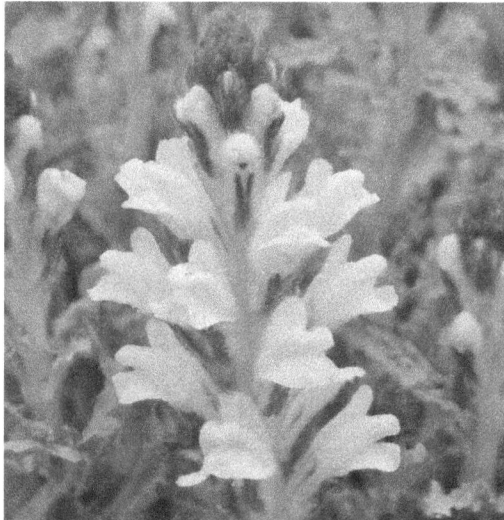

The broomrape, scientifically, the *Orobanchaceae,* are mostly parasitic plants with over 2,000 species worldwide. The members of this family are plants like broomrapes, paintbrushes, warriors' plumes, cream sacs, owls, clover, etc.

The identifying features of plants in the broomrape family are:

- They are annual, perennial herbs or shrubs that grow in temperate habitats and tropical Africa.

- The leaves are simple, alternate in arrangement, and appear like fleshy scales on plants lacking chlorophyll.

- The flowers are in many forms, some modified at the base with 2 fused upper petals and 2 fused lower petals. Refer to the image above.

- The fruits are capsules that are usually dry multichambered fruit with many seeds inside.

24.   Identifying plants in the Horsetail Family

Plants in the horsetail family are also scientifically known as *Equisetaceae*. This family only has 15 species worldwide, but they are not in New Zealand and Australia.

The features of the horsetail family include:

- The plants produce spores and do not have seeds or flowers. It is said that this family is the only surviving family of plants in order of many tree-sized fossils.

- They are perennial herbs growing from horizontal underground stems (rhizome).

- The stems are hollow except at the nodes. Some species have whorls of solid, grooved branches.

- They have incredibly tiny leaves, usually not photosynthetic – refer to the first image.

- They have sac-like spores called sporangia found on the inner surface of the scales on the stem and are wind dispersed.

25.   Identifying plants in the Soapberry plant family

The soapberry family is a family of flowering plants scientifically called *Sapindaceae*. There are about 1,500 species worldwide, with some of the notable species being longan, lychee, pitomba, rambutan, korlan, and ackee.

Most species grow in tropical and subtropical habitats.

The notable features of plants in the soapberry family include:

- They are deciduous trees and shrubs (sheds leaves annually), and some are woody vines.

- The leaves are generally compound and arranged alternately.

- The flowers are small, with prominent nectar disks between the petals and stamens.

- The fruits are schizocarp, dried fruits with 2 single-seeded segments – can also be single-seeded capsules.

26.   Identifying Plants of the Primrose Family

The primrose family is a flowering plant family also called, scientifically, the *Primulaceae.*

This plant family has around 600 species around the world. Some of the most notable species include; creeping jenny, florists cyclamen, scarlet pimpernel, and yellow loosestrife.

Some of the notable features of the primrose family include:

- Simple leaves that may be basal rosette or may be opposite or whorled.

- They are annual or perennial herbs or even woody plants.

- The flowers are umbel, that is, flowers with short stalks coming from a single point.

- The fruits are full capsules.

- The primrose family is not related to the evening primrose family.

27. Identifying Plants in the Figwort Family

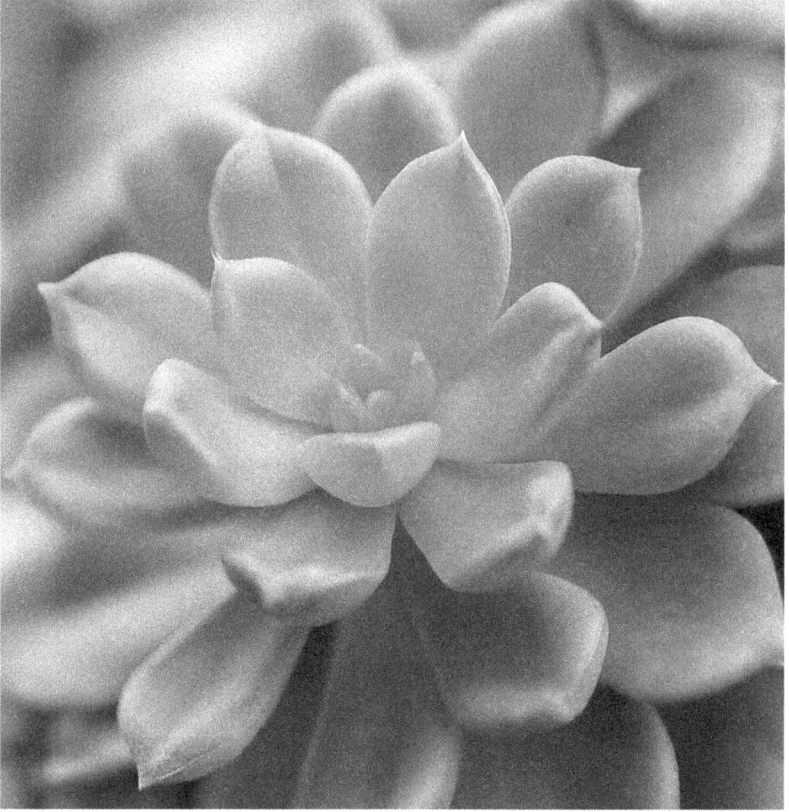

The figwort family is scientifically called the *Scrophulariaceae* family. It is a family of flowering plants with around 1,700 species across the world. Some of its notable species are the bee plant and the butterfly bush.

Some notable features of the figwort family include:

- They are annuals or herbaceous perennials, shrubs, and trees.

- The leaves are simple and generally full, with an alternating flower arrangement.

- The flowers are in many forms and usually have modified leaves at the base.

- The flowers have 5 fused sepals and 4 to 5 fused petals.

- The fruits are a capsule.

28.　　Identifying Plants in the Silk Tassel Family

The silk tassel is scientifically called *Garryaceae,* with only 14 species primarily found in temperate and -sub-tropical habitats.

Some of the iconic features of plants in the silk tassel family include:

- They are evergreen shrubs and small trees growing in temperate and subtropical habitats

- The leaves are simply arranged opposite each other along the stem, usually dark grey, green grey, ovate and leathery.

- The fruits are usually berries containing multiple seeds inside.

29. Identifying Plants in the Butchers Broom Family

The interesting name butcher broom came from ancient times when the stiff branches were put together in a bundle to make brooms that cleaned the butchers' cutting blocks.

Scientifically the butcher's broom family is called *Ruscaceae*. There are about 475 species worldwide; examples include Solomon's seal and bear grass.

Some of the iconic features of plants in the butcher's broom family include:

- They are perennial herbs and shrubs that grow from seeds or horizontal underground stems(rhizomes)

- The leaves are basal or stem with sheathing on the stems or sometimes small scales on the stems.

- The flowers are in clusters borne from branchlets or on one side of the branchlet. The flowers can also be an unbranched stem with stalked flowers that open from the bottom going up.

- The fruits are of two types, some being berries that are multi-seeded with a fleshy ovary wall and others are papery capsules, ie., dry multichambered fruit that splits when mature.

30.   Identifying Fungi- Mushroom Species

Mushrooms are one of the many species of fungi called *Basidiomycota*. There are about 14,4000 species of fungi, including yeast, rust, molds, and mushrooms.

Fungi were initially part of the plant kingdom but have been separated because they lack chlorophyll like the rest of the plants; however, many fungi are useful to plants and humans in that, together with bacteria, they can breakdown organic matter into oxygen, carbon dioxide, nitrogen and phosphorus in the soil and into the atmosphere. This time we will only focus on mushrooms.

Mushrooms are somewhat like the fruitbody of the fungus, appearing above the ground or on its food source. Toadstool is the poisonous version of mushrooms which can be fatal. Do not eat mushrooms unless you are very sure about them.

Here is how to identify edible mushrooms from poisonous ones.

- Look for the ones with brown or tan gills. The ones with white gills represent some poisonous mushrooms. There are mushrooms with edible white gills, but avoid them all to be safe.

- Avoid mushrooms with red caps or stems because most are poisonous; instead, look for the ones with white or brown caps and stems.

- Avoid mushrooms with scaly darker spots because those spots tend to be common among the poisonous varieties of mushrooms.

- Skip mushrooms with a veil-like ring of tissues under the cap because that feature is mostly found in poisonous mushrooms.

In the next chapter, we shall look at some of the individual edible mushrooms and their benefits.

Chapter 5: Edible, Medicinal, and Other Useful Plants

One of the reasons people go wildcrafting is to search for edible, medicinal plants and ones that can be used for cosmetic purposes. In the previous chapter, we looked at how well to identify different plants.

In this chapter, we shall be looking deeper into what plants give you what value so that when you are out there foraging, you will know the value of the different plants.

Edible plants are grouped according to the part of the plant that is eaten. Some plants fit into more than one group. For example, in pumpkins, both the leaves and the fruits are edible. The edible families can be used as spices, fruits, vegetables, or herbs.

Kindly ensure you add the plant's nutrients so that you can know what you are getting when you use a plant. Is it rich in carbohydrates, proteins, vitamins (which ones), micronutrients, antioxidants, etc.?

If a plant can treat something, maybe mention its properties that help with that.

Plant	Image	Edible Part and Uses
African Olives		African Olives are in the Oleaceae plant family and are closely related to the commercial olive. Olive fruits are pickled or cured with water, brine, salt, oil, or lye. They are then eaten as a relish or used in bread, soup, or salads. When dried in the sun, olives can be eaten without curing. African olives contain vitamin E, and K, and antioxidants such as oleuropein, hydroxytyrosol, and tyrosol which has anti-cancer effects, and oleanolic acid which prevents liver damage and reduces inflammation.
Alfafa		Alfafa is in the Fabaceae (pea) or Leguminosae plant family. It is grown by farmers for pasturage, however, its leaves and young shoot can be eaten

raw or cooked.

The leaves can also be dried to be used as a tea or in soup. Alfafa is to be used moderately by patients with systematic lupus erythematosus as it can trigger attacks.

Alfafa seeds are commonly used in salads, and soups, eaten raw or added to cereal flour when dried and ground, for their essential vitamins and minerals. Alfafa is rich in vitamins A, C, E, and K, and minerals like calcium, iron, phosphorous, and potassium.

Alsike Clover

Alsike clover is in the Fabaceae (pea) plant family and grows along the roadside and waste places.

It has edible leaves and flower heads which can be cooked or consumed raw. Its dried flowers are used to make healthy tea or ground to make flour.

Alsike leaves are rich in

		iron, Vitamin C, and fiber.
Angelica		Angelica is a wild edible plant in the Apiaceae plant family. The stems are used in salads for their licorice flavor and their aroma, while their stalks and shoots can be cooked or eaten raw without peeling.

The leaves, roots, or seeds can be used to make tea. The roots are used to relieve menstrual cramps by warming, relaxing, decongesting, and stimulating blood flow.

Angelica roots are a good source of Vitamin B12, thiamin, magnesium, iron, potassium, and riboflavin.

Angelica is used for its decongesting effect to help ease indigestion, gas, and bloating. |

Arctic Rasberry		Arctic raspberries are in the Rose family and are also known as arctic plumboy, nagoon berry, or arctic bramble.
		The berries are normally used as a base for wines and liquors, can be eaten raw or baked, or can be used to make jam, juice, and nectar.
		The leaves (dry or fresh) and the flowers, can be tossed into a salad or used as tea.
		Arctic raspberry is a good source of vitamins C and K, antioxidants, and essential fatty acids.
Amaranth species		Amaranth species belong to the Amaranthaceae plant family or simply the amaranth family.
		Amaranth is a celebrated food around the world, from cultivation to wild harvests of its grains and leaves.
		The seeds of the amaranth species are a good source of proteins, lysin which is an

		essential amino acid, and vitamins C, E, potassium, iron, magnesium, and phosphorus. They are crushed and heated first before eating.

The leaves of amaranth are harvested green and cooked like spinach, and can also be used in smoothies. The leaves are a good source of fiber, iron, and vitamin c. |
| Arrowhead |  | Arrowhead is an aquatic plant in the Alismataceae plant family. Their common name is duck potato.

The edible part of arrowheads is the tubers which are either roasted or boiled before they are eaten.

The tubers are a good source of starch and phosphorous. |
| Asian mustard |  | Asian mustard also called African mustard belongs to the Brassicaceae plant family.

The leaves and shoots are consumed for their rich |

nutrients. They are low in calories, and rich in fiber, magnesium, vitamin K and other micronutrients.

Beach mustard

Beach mustard is a green in the Brassicaceae or mustard plant family. It grows on the beaches of most coastal areas and in sand dunes.

The entire beach mustard plant is edible, with the leaves being served raw as a peppery garnish because of their fleshy salty taste, and the younger leaves are used in salads.

The roots can be dried and ground into a powder that can be used to make bread.

Beach Mustard is a good source of selenium and magnesium, iron, and manganese copper and is rich in zinc and calcium. The leaves are also a good source of fiber.

Blackberries	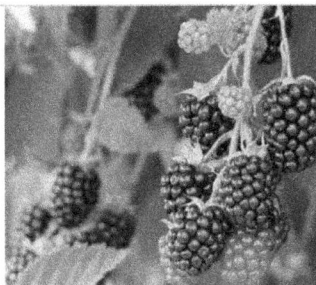	Blackberries are in the Rose family or Rosaceae, a family of pricky fruit-bearing bushes. Blackberries can be eaten raw or used as cake and cookie toppings. Blackberries have several medicinal and nutritional benefits. They are a good source of vitamin C and can help treat wounds, regenerate skin and prevent scurvy.
Ball mustard	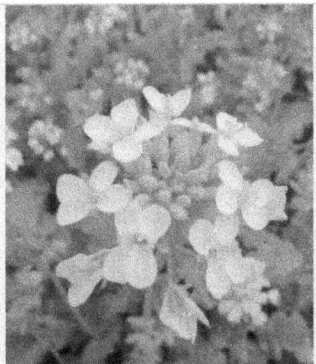	Ball mustard is in the Brassicaceae plant family. The leaves can be consumed raw or cooked while the younger leaves can be used as a flavoring in salads. Flowers can be consumed as vegetables or used as a garnish in salads. The seeds are a source of edible oil. Nutritionally, ball mustard is rich in iron, fiber, vitamin C and E .

Beggerticks

Beggersticks are also called the blackjack, Bidens pilosa, hitchhikers, Spanish needle, cobblers peg, or farmers friends and belong to the Asteraceae plant family.

The flowers of beggarticks can be candied or tossed in a salad.

The infusions of beggarticks can be used as an herbal remedy for inflammation, irritation, and pain.

Beggarticks contain vitamin C, which acts as an antioxidant and also helps the body make collagen which helps heal wounds.

Billberry

Billberry is also known as European berry, whortleberry, or huckleberry, and it is in the heather family/Ericaceae.

Bilberries have been widely cultivated, although it is also commonly found in the wild.

The berries of this shrub are

		edible and used in jams, ice cream, or pies. The berries can also be cooked or eaten raw. The leaves and the berries can be dried and used as tea. Bilberries are rich in vitamin C and were traditionally used to cure scurvy.
Bluebeard lily	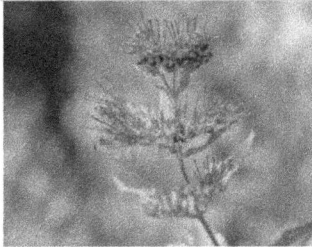	Bluebeard lily is in the Liliaceae plant family and is also commonly known as clintonia, Clinton's lily, snakeberry, cow tongue, dogberry, or corn lily. Its young leaves can be eaten raw or can be cooked. The leaves have disinfectant and sedative properties. Note: Do not consume any other part of the blue bead lily because they can be poisonous

Broadleaf plantain

Broadleaf plantain is in the Plantaginaceae family that is not just good for good health but as a herbal medicine to treat chronic diarrhea and digestive tract disorders.

The entire broadleaf plantain plant is edible, with the leaves being eaten raw or cooked. The leaves can be used in salads but are usually first boiled to make them tender.

The seeds are ground and mixed with flour, while the leaves can be dried and used as herbal tea.

Broadleaf plantain has microbial and antioxidant attributes and anti-inflammatory and anti-tumor properties. It also improves the immune system and treats ulcers, diarrhea, and fatigue.

Bull Thistle		Bull thistle is in the Asteraceae family, and like other members of this family, it has sharp prickles that make it difficult to touch. It is also commonly called cotton thistle. The roots of bull thistle can are best mixed with other vegetables. They are rich in inulin (a complex sugar present in the roots of some plants). The flower stems can be cooked while its young leaves can be used in making salads. Ensure to remove prickles before eating the leaves to avoid injuries. Bull thistle roots can be roasted.
Bitter dock		Bitter dock is a member of Polygonaceae or the buckwheat plant family. The leaves of the bitter dock can be eaten raw or cooked, although they have a bitter taste. Because of this, some people prefer to boil and

change the water at least once to reduce the bitterness.

Young stems of the bitter dock can also be cooked and eaten.

The seeds can be dried, ground into powder, and then used in baking.

The bitter dock is a source of thiamin, niacin, folate, calcium, dietary fiber, vitamins A, C, and minerals like iron, magnesium, and phosphorus.

Chickweed

Chickweed is in the Caryophyllaceae family. They have edible leaves that are eaten raw and are best added to salads and sandwiches.

The stems and flowers can also be added to food. It is rich in vitamins A, D, B complex, and C, calcium, potassium, zinc, manganese, sodium, copper, and iron.

Canyon Grape		Canyon grape is a type of wild grape growing on deciduous vines commonly known as Arizona grape or uva del monte, and it's in the Vitaceae family.
		Ripe canyon grapes are sweet fruits that are used to make juice, syrups, sweets, and wine.
		The leaves can be used in soup and stew. Young tendrils can be picked or tossed in a salad.
		Nutritionally, Canyon grape is rich in vitamin C, potassium, and vitamin K.
Creeping Charlie		Creeping charlie is commonly known as ground Ivy, over the ground, alehoof, field balm, runaway robin, or coltsfoot. Creeping charlie is an aromatic weed like other mint family plants.
		The leaves have a mild mint-like flavor and are therefore used in salads.
		The plant can also be dried

and used in tea or beer to add flavor. The leaves also can be cooked like spinach or added to stews and soup. They are high in vitamin C and are used to treat scurvy.

Crimson Clover

Crimson clover, scarlet clover, or Italian clover is in the Fabaceae plant family and is a wild plant, although it is also cultivated.

The seed and flowers are edible, with the seeds being sprouted and used in salads or being dried and ground into flour while the flowers are dried or used fresh as tea.

Crimson Clover is rich in vitamin C, which helps to boost the immune system and stimulates the production of white blood cells.

Crowberry		Crowberry belongs to the heather family and is also known as moss berry or generically as Empetrum. Their ripe berries are the only edible part of the plant and are used to make wine, pies, jams, and jellies. The ripe berries can also be eaten raw and are a good source of fiber, antioxidants, vitamin C, and vitamin K.
Coontail		Coontail is an aquatic plant in the Ceratophyllaceae family and is also known as hornwort or rigid hornwort. Coontail is a good cover for fish and a habitat for aquatic microorganisms. Coontail is medicinal and can be used to treat dermatitis, fever, sunburn, and scorpion stings. It is used as a health supplement because of its low calories and its proteins, magnesium, and calcium.

Cleavers		Cleavers are in the Rubiaceae plant family, also called sticky willy, catchweed, or clivers.
		It is hard to eat cleavers because they are stingy and bitter, yet they are very beneficial in that they are rich in high-quality vitamin C and minerals like silica needed for healthy nails, hair, and teeth.
		The seeds are edible when dried and lightly roasted; their young shoots can be eaten when cooked, while the whole plant, when dried, can be used as tea.
		Cleavers are used to cure ulcers and skin problems.
Daisy fleabane		Daisy fleabane is in the Asteraceae family.
		Only its leaves are edible and are only pleasant if cooked with greens.
		The plant can also be used to treat digestive ailments.
		Daisy fleabane has a high

		vitamin C content and iron.
Dandelion		Dandelions are in the family of Asteraceae and are also known as lions' teeth. The plants have edible roots, leaves, and flowers.

The leaves can be used in salads or cooked, and the flowers are used in juices. The roots can be made into a coffee substitute or, together with the leaves, can be dried and made into tea.

Dandelion leaves are an excellent source of vitamins A, C, K, and E. They also contain folate and small amounts of vitamin B. |
| Downy Yellow violet | | Downy Yellow Violet is in the Violaceae plant family and is also known as yellow violet.

The leaves and flowers are edible and are high in vitamins A and C, while the roots can be used for medicinal purposes to cleanse the blood, as a respiratory remedy, and |

lymphatic stimulant.

Yellow dock

Yellow dock is in the Polygonaceae family, the same as bitter dock, and is also known as curly dock.

The leaves can be eaten in small quantities until flowers begin to appear. Leaves can be eaten both raw or cooked. The stems are eaten raw or cooked but make sure to peel first. The seeds can be eaten raw or cooked and can be dried to be used as a coffee substitute.

Yellow dock is rich in vitamins A and C, which are also antioxidants.

The roots are medicinal since you can boil them to make detox tea that helps clean the liver and clear skin ailments.

Elecampane		Elecampane is in the sunflower (Asteraceae) plant family and is also known as horse heal or elf dock. The leaves are edible and better cooked because the raw version is bitter. The roots are also edible but contain inulin which may cause gas; however, they can be used as a natural food flavoring. Elecampane is rich in inulin and other polysaccharides like mucilage and sterols. For that, it is used to treat respiratory ailments and diabetes.
Evening Primrose		Evening Primrose is in the Onagraceae plant family. The roots are fleshy and sweet in the first year; the shoots are also edible. The flowers can be used in salads. The plant has an anti-inflammatory effect and is known to reduce pain and

improve mobility.

Evening primrose is used to make primrose oil from its seeds that treat brittle hair and nails and treats many health ailments.

Evening primrose oil contains omega-6 and essential fatty acids

| Fennel | | Fennel is in the Apiaceae plant family and has been widely cultivated, although it remains a wild plant.

Fennel is edible from its roots to its seeds. The leaves are best eaten young, while the seeds are aromatic and are therefore used in flavoring cakes and stuffing meat.

Fennel contains beta carotene, which is converted to vitamins in the body, and vitamin C, which is essential for producing collagen, which helps repair tissues in the body. |

Forget me not		Forget me not is in the Boraginaceae family, also known as scorpion grasses. Their flowers are the edible part of the plant and can be eaten raw, used in salads, for decorating cupcakes, or as garnish. Forget me not reduces high blood pressure, soothes the nerves, and promotes a restful night's sleep.
False Solomon Seal		False Solomon's seal is in the Asparagaceae and lily plant family, also known as false spikenard or Solomons plume. Their berries are bittersweet and are edible together with their young leaves. Traditionally, their rhizomes were dried to brew tea that was used to treat coughs and lung disorders NOTE: Never eat any part of the true Solomon seal, which is a close look-alike to the false Solomon seal, because it is poisonous.

Goldenrod		Goldenrod is in the sunflower or the Asteraceae plant family. The flowers can be eaten or used as a garnish in salads, and also, the flowers, together with the leaves, when dried, can be used to make tea. The leaves can be cooked or added to stew and soup. Goldenrod has beneficial plant compounds like quercetin, flavonoid, and kaempferol that can hinder the growth of harmful bacteria and yeast.
Ground Elder		Ground elder is in the carrot/Apiaceae family and is also known as bishops weed, goutweed or goutwort, snow in the mountain, or wild masterwort. The leaves can be eaten raw or cooked, and the leaves are best for consumption before the plant flowers. The leaves can also be used

		in salads, soups, or stews. Ground elder contains vitamins C and A and minerals like calcium, potassium, and silicic acid. All these help to alkalize the body system.
Hibiscus		Hibiscus is in the mallow family and is commonly known as rose mallow or rose of Sharon. Young hibiscus leaves can be eaten raw or cooked. The leaves are also used to make tea. The stems are used to make cordage and paper. The hibiscus plant is rich in vitamin C (ascorbic acid), an antioxidant that helps boost the immune system. It can also be used to treat the skin. The antioxidants present, called anthocyanins, helps to heal the skin. The leaves, stems, and roots are used to make dyes.

Horsetail is in the Equisetaceae plant family and has been used for centuries to aid conditions such as arthritis and kidney problems.

This is because the plant contains medicinal properties such as antibiotics, antiseptic, and anti-hemorrhagic.

The stem and leaves can be dried to make tea or cooked and added to stews and soups.

Kudzu

Kudzu is in the Fabaceae plant family and acts as a ground cover and a livestock forage.

Kudzu leaves, flowers, vines, and roots are edible and are used to make syrups and candies. The flowers can be used in salads or cooked, while the roots can also be used to thicken the soup. The leaves can be consumed raw.

Kudzu has antioxidant properties. It is especially

		used in Asian countries to treat alcoholism, ease menopause symptoms and reduce inflammation.
Lavender		Lavender is in the Lamiaceae plant family and produces purple flowers. Though a widely cultivated plant for its camphor, flavor, and scent, it is also common in the wild. Lavender can produce pinkish dye and be added to different products for its scent. Lavender can also be used in place of Rosemary to cook. It is also used to produce lavender oil, which has antiseptic and anti-inflammatory properties and therefore helps to heal minor burns and bug bites. The scent is also used to reduce anxiety, depression, insomnia, and restlessness.

| Miner's lettuce | 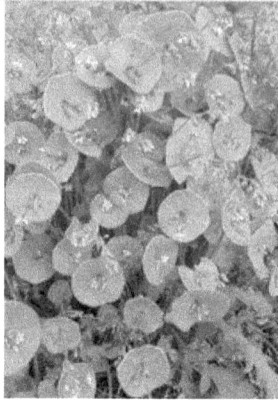 | Miner's lettuce is in the Moniliaceae plant family and is known for its good taste. Apart from the taste, it's a good source of vitamin C, vitamin A, and iron.

The leaves, flowers, and roots of Miner's lettuce are edible, with the flowers being used in salads and leaves being cooked or eaten raw. Older leaves can be bitter. |
| Mushrooms |  | Mushrooms come in different species, both edible and poisonous. Below we will look at some of the edible species of mushrooms.

Mushrooms are a good source of fiber, proteins, and antioxidants. |

Flowery Blewit		Flowery Blewit is a short, medium-sized mushroom with a flowery odor. It is best eaten when fully cooked.
Giant PuffBall	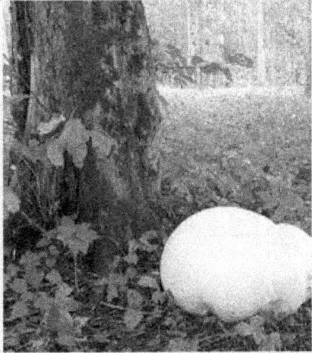	Typically giant puffball grows from 4 inches to 27 inches and is an excellent source of food. It has a pleasant scent and a faint taste. It needs to be fully cooked before it is eaten. Also, before cooking, remove the outer skin by peeling it, do not wash it with water because it will become soggy and will be wasted.

Reishi		Reishi is a type of mushroom that grows on dead wood. When growing, it starts with a white margin that changes color over time into shiny lacquered skin, but inside it remains to have white flesh.

Reishi is edible when broken into smaller pieces and used in meals, desserts, and smoothies. |
| Oyster mushrooms | | Oyster mushroom is parasitic and grows on trees. They have white to dark brown caps with whiteish-yellow gills.

Oyster mushrooms can be eaten raw or can be cooked, but it is best cooked to allow proper digestion. |
| Common Morel | | Common morel starts from pale cream to yellowish brown and darkens as it grows older.

Common morel has a pit-like network of ridges that make it look like a sponge, |

and it is edible. However, there are many false morels out there, so if you are not sure, you should skip them because false morels are poisonous.

False morels have wrinkled caps that fall over the stem.

Common morels are best eaten cooked especially if added to other meals.

Chapter 6: Toxic Plants

Plants are identified as toxic or poisonous because one or more of their parts could cause a harmful reaction if inhaled, consumed or if you came into contact with them.

In the previous chapter, we learned how to identify most plants. In this chapter, we will focus on learning to identify toxic plants and some safety tips on what to do in case you accidentally consume or come into contact with these plants.

Note: Some plants have both edible and poisonous parts, and you need to be aware of this.

Poisonous plants

Watch out for the **Parsley Family** *(Apiaceae)*

This is also called the carrot family, with plants that can be easily confused with medicinal plants. In particular, two plants in this family, Poison Hemlock and water hemlock can kill you or your livestock.

Let's look at these plants' features and their similarities with wild carrots:

Carrot	Poison Hemlock	Water Hemlock
The part of a carrot that is edible is the taproot but the leaves are also edible.	All parts of Poison hemlock are poisonous.	All parts of the water hemlock are poisonous, but the roots have the highest poison concentration.
Carrots come in yellow, red, white, and purple colors.	Poison hemlock has white roots.	The water hemlock has white roots.
The stems are umbels (a cluster of flowers with stalks of nearly equal length springing from the same point) with many other umbels.	Stems have reddish-purple spots are hollow, and are not hairy	The branches have thick root stalks that store a brown or straw-colored highly poisonous liquid released when the stem is broken.
Leaves are compound and green and grow alternately along the stem.	The leaves are bright green, fernlike, and have a strong, musty odor when mushed.	The leaves are double or triple compound with large leaflets compared to carrots and the poison hemlock.

Carrots have small flowers appearing in umbrella-like clusters.	Flowers are small and white, arranged in clusters at the end of the stem.	They have small white umbrella-like flowers growing in clusters.
The roots in carrots have color pigmentation called beta-carotene, which, when eaten by humans, is converted into vitamin A, which is good for healthy eyes, bones, teeth, and skin.	The roots are long white taproot that is highly fibrous.	The roots are thick fleshy slender tubers growing below the rootstalk.

When you know their differences, it will be hard for you to confuse them.

Other Poisonous Plants

1. Poison oak, Poison Sumac and Poison Ivy

Poison Ivy plant

Poison Ivy belongs to a plant family called Anacardiaceae, which is the same family as mangoes and cashews. What these plants have in common is that they produce urushiol. This is an oily sap which leads to contact dermatitis when it touches 90% of human skin. The skin rash can last up to a week and then disappear. You can treat contact dermatitis with over-the-counter medications like anti-itch creams.

Note: Every part of the poison Ivy plant is poisonous, especially when burnt. The smoke gets volatilized, and if it gets to the lungs when inhaled could cause death.

2. Stinging Nettle

Stinging nettle is in the Urticaceae family and has simple leaves that are oppositely arranged. They grow along streams and trials and grow in colonies.

Stinging nettle has a lot of nutritional and medicinal benefits, yet it is toxic to the skin. For you to harvest stinging nettle, you will need to wear something to protect your skin, like gloves, because you should not touch it.

Its sting is worse than that of a bee, sharp and excruciating, causing a skin rash lasting up to 12 hours.

The sting is a microscopic hollow needle from the nettle plant that injects histamine, acetylcholine, serotonin, and formic acid. These cause the sting, burning, and painful reactions to the skin.

This aside, stinging nettle leaves are very nutritious and are used as food, the stems are a source of fiber and are used to make clothes, and various extracts from different parts of the plant are used as medicine.

3. Baneberries

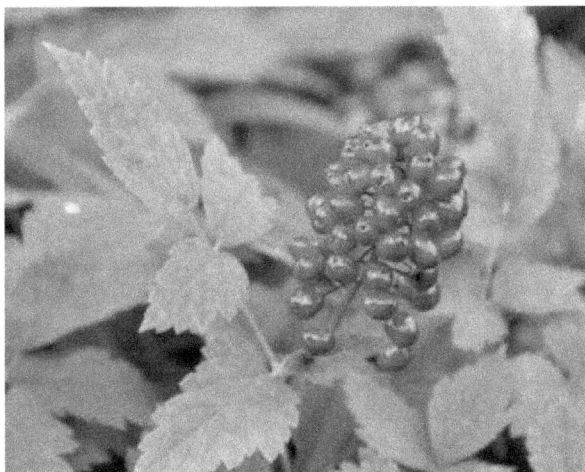

Baneberries are from a family of plants called the buttercup family, baneberry being the common name representing several species. This group of plants produces very attractive berries but are very poisonous.

These plants have compound leaves and one flower stalk originating from a single stem. The leaves are lobed with toothed leaflets that have hairy veins on them. There are white and red baneberries that have identical leaves.

The flowers are white and appear in clusters; they are slightly fragrant but lack nectar.

The fruit berries start green and grow into attractive white or red berries with a black spot at the tip.

The entire baneberry plant is toxic, with the poison being most concentrated in the roots and berries. If consumed, it could lead to cardiac arrest. The good news is that baneberries are very bitter, so the chances of being consumed by humans are low.

Despite how poisonous baneberry is, you can still use it for medicinal use to treat colds and skin conditions. But you should not use it raw, especially when you do not know how to make the medicine.

4. Bittersweet Night Shade

Bittersweet nightshade comes from a plant family called *Solanaceae*, a family of flowering annuals and perennial herbs, vines, epiphytes, shrubs, lianas, and trees. Other plants like potatoes, tomatoes, eggplant, bell, and chili pepper also belong to this family.

You can notice that the leaves of bittersweet nightshade and those of potatoes are very similar in appearance.

Their leaves are dark green to purplish with one or two lobes near the base, while the flowers are star-like, purple, and with a yellow cone. The flowers grow from the stalk extending from the stems. When the leaves are crushed, they have an unpleasant smell.

Bittersweet nightshade is a poisonous plant to both humans and animals. The good thing is that the plant has an unpleasant smell, so animals can sense it and avoid eating the plant.

Rare cases of death have been reported from this plant.

Members of the nightshade family have a toxin called solanine and a glycoside called dulcamarine. The berries and leaves have the most concentration of toxins, and the number of toxins in the plants also varies with the soil where the plants are growing, the stage of growth, and the climate.

Even though this plant is somewhat poisonous, the stem of the bittersweet nightshade can be used to make medicine to treat skin conditions such as eczema, acne, boils, broken skin, and warts. It is best to leave herbal experts to use such plants.

5. Castor Bean

Castor bean is a seed from the castor oil plant, originating in East Africa but currently found worldwide. The castor oil plant is a flowering plant in the spurge family, scientifically called the *Euphorbiaceae*. The plant has glossy leaves with 5 to 12 lobes with coarsely toothed segments.

Castor seeds are castor beans that are not beans but are given the name for their -bean-like appearance. Castor beans contain a poisonous toxin known as ricin, which, if eaten raw, will block cells from making proteins needed by the cells to function. If cells in living organisms, including humans, do not function, it leads to death, and even the tiniest amount of ricin can cause death.

Castor oil is one of the products that is made from castor beans and is very useful. However, it is not poisonous, and it does not contain ricin because it is made from castor seeds that have been cooled, dried, and pressed, which removes the ricin.

Poisoning from castor seeds is rare, and for it to happen, the bean must be chewed before swallowing, leading to ricin being released.

6. Dumb Cane

Dumb cane or the Leopard Lily is a flowering plant used as a house plant in many cases and belongs to the Araceae family, which is widely cultivated as an ornamental plant.

The plants have a straight stem and simple, alternate leaves with white spots that make them ornamental.

Do you wonder why this name or where it came from? Well, chewing the leaves of a dumb cane is known to cause temporary speechlessness. The cells of the plants have stinging crystals called raphides, containing calcium oxalate, which if chewed or ingested, irritates the mucus membrane, causing inflammation of the lips, mouth, tongue, and throat. This can continue for a few days.

The effects of dumb cane are not life-threatening and can be treated.

7. Calla Lily

Calla Lily is a common houseplant from the Araceae family, the same as the dumb cane above. Calla lily is a flowering plant with flowers in yellow, pink, orange, rose, lavender, and dark maroon colors. Other common names are arum lily or pig lily.

These plants are very attractive, as in the image above, and are cultivated for ornamental purposes, but you may find kids accidentally chewing the leaves or flowers.

All parts of the calla lily contain calcium oxalate crystals released from the plant if chewed or bitten. Usually, symptoms like pain and burning of the lips, mouth, tongue, gums, and throat occur immediately. It may also cause nausea and vomiting. The symptoms are not life-threatening and can be treated.

If someone consumes the calla lily, do not force them to vomit; remove the parts of the plant in the mouth, wipe the tongue with a wet cloth, and rinse the lips with water. Give them/drink milk or creamy snacks like yogurt, ice cream, or popsicle to relieve the pain. If the person has swellings on the tongue or throat, do not give them something to eat; rather, ensure they get medical attention.

What if someone consumed a toxic plant? What to do

- Stay calm, and do not panic.

- Call for emergency assistance. Make sure you know the emergency number for your country. You can also call the poison control center for your country.

- If you are out somewhere far from medical facilities and people, use activated charcoal if you have it. Then seek emergency help immediately.

- Do not administer any medicine without a prescription, and do not attempt to force vomit because this could make things worse.

Before harvesting anywhere, ensure that there is no contamination. Keep in mind that areas near roads, farms and near industries are most likely to be polluted with pesticides and chemicals.

Conclusion

Nature has a lot to offer, from free food to free medicinal resources out there in the wild waiting for you to discover them, and as long as you have put your heart into foraging, you will enjoy it immensely.

But beware of those toxic poisonous plants and you are good to go.

Happy foraging!

www.ingramcontent.com/pod-product-compliance
Lightning Source LLC
Chambersburg PA
CBHW032229080426
42735CB00008B/780